YOUR KNOWLEDGE HAS VALUE

- We will publish your bachelor's and
  master's thesis, essays and papers

- Your own eBook and book -
  sold worldwide in all relevant shops

- Earn money with each sale

Upload your text at www.GRIN.com
and publish for free

Autonomous Driving. History Vehicles, Technical Specifications/Classifications and the Trolley Effect

Stefanie Öllerer

Bibliographic information published by the German National Library:

The German National Library lists this publication in the National Bibliography; detailed bibliographic data are available on the Internet at http://dnb.dnb.de.

ISBN: 9783346889379
This book is also available as an ebook.

© GRIN Publishing GmbH
Trappentreustraße 1
80339 München

Print and binding: Books on Demand GmbH, Norderstedt, Germany
Printed on acid-free paper from responsible sources.

GRIN web shop: https://www.grin.com/document/1363671

Table of Content

1. Introduction

The whole world is constantly evolving. Before the 17th century, mechanics and electricity were not invented for mankind (Hery-Moßmann 2022). Many discoveries and inventions have been an enrichment for the human life in the last 300 years, but also many are gratuitous or even endangered (Pogemann 2019). New technologies are not only available to assist people's activities or to produce commodities, but they are and will be of much more value in the future (Tremmel 2020). For the first ten years of the 21st century, the masses had been guaranteed unimaginable technical advances and the continuing possibilities of self thinking processes in metal (Bosh Global 2018). It is named Artificial Intelligence, in short AI. But this engineering marvel is not yet fully functional (Maione 2022). The machine cannot think on its own so an ethnic decision comes from the developer (Maione 2022). However "Artificial Intelligence is the key for autonomous driving" (Pink 2018) and revolutionized the transport sector in all aspects (Bosh Global 2018). This journey to a complete automated infrastructure, whose development is beyond all questions, will affect our lives (Hammele and Kuhnert 2022). An enormous number of experts will be required for the designing and upkeeping of the new cars (Tyborsky 2020).

I am currently doing a training as an IT-Specialist. For me it is a wonderful feeling to program some codes and see how the machine will perform. I am so thankful that I found a job. Yes, I want to do a job that is hopefully needed for the future and perhaps impacts the future. So, for me and for thousands of other young people who are interested in science and technology, this revoulution would be a big chance.

2. History of Vehicles

Around five to seven million years ago, the first human appeared and began walking on two legs (Noble Wilford 2002). After that people found ways to move other than just by their foot.

This image has been removed due to copyright issues.

Illustration 1: Horses which pull a cariage (Looser 2022)

In the bigger part of the human history, the fastest way to move over land on a horse was discovered. Horses have been of great importance especially in carrying people around. "They can run very fast and very far." (American Museum of Natural History 2020) Furthermore, these animals eat grass, so they can go everywhere people also can. The digestive system of a horse allows them to walk all day. Horses therefore changed the history, by transporting people between civilizations. (American Museum of Natural History 2020)

After humans traveled with horses the steam car was invented by Nicholas Joseph Cugnot in 1769 (Lienhard 2000).

This image has been removed due to copyright issues.

Illustration 2: The Steam Car (Burgess Glasscock 2011)

This car in Illustration 2 was the first self-propelled vehicle built for road. But it was really used as an artillery tractor. First it was demonstrated to the government officials in Paris, but it was never a mode of transport for the wide public. It ran with a tricycle layout, two rear wheels and a steerable front weel controlled by a tiller. The vehicle could also speed up to 2 miles per hour in 15 minutes. (Burgess Glasscock 2011)

This image has been removed due to copyright issues.

Illustration 3: The First Car (Woodford 2022)

The Illustration 3 shows the first car, which was a motorwagon made in 1855 by Karl and Bertha Benz. It had three different sized wheels in a cylinder engine. With the gas power, it could reach speeds up to 10 miles per hour. Additionally, an internal combustion engine was added along with gears to go uphill as well as turn the vehicle. (Woodford 2022)

This image has been removed due to copyright issues.

Illustration 4: Ford Model T (Meriggi 2018)

The Ford Motor Company invented the Model T (Illustration 4) in 1908. The first affordable automobile was made by assembly line productions. It was also the first global car, which was sold over 20 years (in total 5 million pieces). Cars could then reach speeds of up to 40-45 miles per hour with an engine of 20 horsepower. (History.com Editiors 2019)

The Trabant in Illustration 5 is an automobile, which was produced from 1957 to 1991 by former Eas German car manufacturer VEB Sachsenring Automobilwerke Zwickkau (Miltimore 2019).

This image has been removed due to copyright issues.

Illustration 5: The Trabant (Greuel 2018)

This car had unusual features at that time (a transverse engine, a hard plastic body mounted on a one-piece steel chassis, front-wheel drive, and independent suspension). It is often seen as a symbol of the defunct East Germany. (Greuel 2018)

Now we have a range of different cars that we can buy with a lot of money. In 2022, the best selling car in Germany was the VW Golf (Illustration 6) (Bekker 2023).

This image has been removed due to copyright issues.

Illustration 6: VW Golf (VW Golf AG 2022)

2.1 Technical Classifications

Autonomous driving is not only developed on one level. There are different levels in car ranging from 0 to 5. (Faist 2023)

This image has been removed due to copyright issues.

Illustration 7: Levels of Autonomous Driving (Team Ackodrive 2022)

Level 0 (No Automation)

"Most vehicles on the road today are level 0: manually controlled" (Snopsys 2023) The driver is a human-being who performs full-time all aspects of the dynamic driving task, even when enhanced by warning or intervention systems (Chai, Nie and Beckder 2020: 70).
In this category the following technologies are found: (Faist 2023)
ABS (Anti-Lock Braking System)
ESP (Electronic Stability Program)
Cruise Control
Blind Spot Warning
Automatic Emergency Braking
Frontal Collision Warning
Lane Departure Warning

Level 1 (Driver Assistance)

The official lowest level of automation is Driver Assistance (Snopsys 2023). A driver assistance system of either steering or acceleration/deceleration is to execute the driving mode-specific. For this, the system uses information about the driving enviroment, but the expectation that the human driver must perfom all remaining spects of the dynamic driving task is still there. (Chai, Nie and Beckder 2020: 70)
These systems are installed in Level 1 vehicles: (Faist 2023)

Electronic Adaptive Speed Regulator

Adaptive Cruise Control

Lane Deeping Assistance

Lane Centering Assistance

Level 2 (Partial Automation)

These cars have advanced driver assistance systems, in short ADAS (Faist 2023). The driving/node-specific execution is available by one ore more driver assistance systems of both steering and acceleration/deceleration. But, the human driver does all the rest dynamic driving tasks. (Chai, Nie and Beckder 2020: 70) Examples of these systems are the Tesla Autopilot and Cadillac (General Motors) Super Cruise systems (Snopsys 2023).

Level 3 (Conditional Automation)

"The jump from Level 2 to Level 3 is substantial from a technological perspective, but subtle if not negligible from a human perspective." (Snopsys 2023) Now there is an automated driving system, which performs all aspects of the dynamic tasks and the human driver will have to respond appropriately to a request to intervene (Chai, Nie and Beckder 2020: 71). Level 3 vehicles are not widespread in the market yet (Faist 2023). But in Europe Audi will rollout the full Level 3 A8L with Traffic Jam Pilot in Germany first. (Snopsys 2023)

Level 4 (High Automation)

"The key difference between Level 3 and Level 4 automation is that Level 4 vehicles can intervene if things go wrong or there is a system failure." (Ackodrive 2022) The driving/node-specific perfomance by an automated driving system carries out all aspects of the dynamic driving tasks, even if a driver does not respond appropriately to a request to intervene (Chai, Nie and Beckder 2020: 71). "However, a human still has the option to manually override." (Snopsys 2023) Currently, there are also buses and tubes, which are in the category of the autonomous driving level 4 (Shea Choksey and Wardlaw 2021).

Level 5 (Full Automation)

Level 5 automation means full-time performance by an automated driving system with all aspects of the dynamic driving task under all roadway and enviromental conditions that can be managed by human driver (Chai, Nie and Beckder 2020: 71). "Fully autonomous cars are undergoing testing in several pockets of the world, but none are yet available to the general public." (Ackodrive 2022)

2.2 The Next Generation

<blockquote>

"The next generation automobile industry

as a creative industry"

(Seio Nakajima 2019: 1)

</blockquote>

"The aim of this paper is to describe the recent transformation of the automobile industry from a manufacturing industry to include aspects of the service and creative industries. Firstly, it reviews the recent trend of the automobile industry as a service industry. Secondly, it discusses the automobile industry's move toward the creative industries. It examines these two trends, mostly based on the next-generation automobile industry in Japan. Finally, it discusses the implications of the above transformation of the automobile industry on academic studies of the creative industries, and argues for what it calls a strong programme in creative industries studies. It also provides a provisional note on government policies in the era of the next-generation automobile industry." (Seio Nakajima 2019: 1)

The abstract of this document from the author is a short overview of the important statement for the next generation and is also the title of the paper. But there are some terms, that must be considered closely to understand the topic.

What is the automobile indusrty exactly?

The automobile industry is the business of producing self-empowered vehicles. This can be passenger cars, trucks, farm equipments and other commercial means of transport. Furthermore, the auto industry has encouraged the development of an extensive road system, made the growth of suburbs possible and shopping centers around major cities. It played a key role in the growth of ancillary industries, such as the oil and travel businesses, by allowing consumers to commute long distances for work, shopping, and entertainment. Also the large number of people that the industry employs has made it a key determinant of economic growth. The car industry has also become one of the largest purchasers of many key industrial products, such as steel. (Infoplease 2022)

What defines the next Generation?

When one searches the internet about the next generation, many different results show up. For example, a film title of Star Trek, named The Next Generation (Fandom 2016) or the articles about Generation Z, which means babys, who were born after 1997 and often referred as the Digital Natives or the iGeneration (Verlinden 2021). But the important point is after a generation comes a next generation and "with every generation comes new expectations and opinions" (Factorial HR 2021).

What does "The next Generation of Automobile Industry" mean?

The next generation of automobile industry deals with the thinking of how consumers engage with their vehicles. For this aspect there are four different observable technology areas, that drives this shift: (GE Digital 2023)

Connectivity – "A connected car is a vehicle which can access the internet through an inbuilt connectivity system." (Team Acko 2023)

Redefined Mobility – The Definition of consumer engagement in a vehicle which changes the very nature of what we consider "mobility" is, by the success of ride-sharing services like Ubers. Most major manufacturers are considering how these trends will impact new development and design. This will result in changes in the physical development of vehicles and the software that is inside them. (GE Digital 2023)

Autonomous Vehicles – "An autonomous vehicle, or a driverless vehicle, is one that is able to operate itself and perform necessary functions without any human intervention, through ability to sense its surroundings." (TWI 2023)

Electrification – Electric vehicles should stop the global warming. They drive only with electricity so during the trip, they do not eject greenhouse gases like carbon dioxide, when the power is clean. Currently, it is the planned technology for the future instead of combustion engine. (Paoli 2022)

What is a creative Industry?

The creative industries are "those industries which have their origin in individual creativity, skill and talent and which have a potential for wealth and job creation through the generation and exploitation of intellectual property." (Parrish 2015) The creative industries include the following terms: (Discover 2019)

Music, performing arts, like acting and visual arts, like painting

Crafts, such as weaving, furniture-making and jewellery-making
Film, TV, animation, visual effects, video, radio and photography
Video games, virtual reality and extended reality
Museums, galleries and heritage, such as stately homes and cathedrals
Publishing and libraries
Design, including product design, graphic design and fashion
Architecture
Advertising and marketing

In Conclusion, the topic in this paper is important as well as a basis e.g. for more discussions (Stevenson 2021).

3. Technical Specifications for Autonomous Driving

There are different and significant specifications (in this coursework there are Radar, LiDAR, Ultrasonic Sensor and Cameras explained) for the functional autonomous vehicle in sensing the enviroment. And nearly all of them make use of the technic of the "Doppler-effect" shown in Illustration 8 (Chai, Nie and Beckder 2020: 17).

This image has been removed due to copyright issues.

Illustration 8: Doppler-Effect (Schmauch 2022)

How does the Doppler-Effect work?

"The Doppler effect, or Doppler shift, describes the changes in frequency of any kind of sound or light wave produced by a moving source with respect to an observer." (Bettex 2010) In addition to sound or light waves, it also works with water waves (Henderson 2015). If the source approaches the observer, waves get compressed by object traveling toward an observer which causes a higher frequency. In contrast, waves emitted by a source traveling away from an observer gets stretched out. (Bettex 2010)

This physical principle is not only used for autonomous driving but also in various other applications: (The Economic Times 2023)

Sirens

Astronomy

Medical imaging and blood flow management

Flow Management

Velocity profile management

Satellite communication

Vibration measurement

Radar – Radio Detection and Ranging

This technique was originally developed for military and avionics applications. They were eventually used to detect hindrances in autonomic vehicles for adaptive cruise control and forward collision warning functions. The Radar sensor make use of the Doppler-Effect and operate in the microwave spectrum (76.6 GHz). The direct energy is emitted from this sensor through the antenna. The emitted energy Pe is reflected by a target, for example an obstacle in front of the sensor, and then is detected by the receiver antenna in the Radar as a received power Pr: (Chai, Nie and Beckder 2020: 18)

$$P_r = \left(\frac{P_e G}{4\pi d^2} \right) \cdot \left(\frac{\sigma}{4\pi d^2} \right) \cdot \left(\frac{G\lambda^2}{4\pi} \right) \cdot (\ )$$

"The first term of this eqation represents the power density of the electromagnetic waves hitting the target. The second term represents the power density of the returning waves. The third term shows the effective aperture or area of the antenna such as rain, snow, or dirt on the antenna […] and the last term represents other influences." (Chai, Nie and Beckder 2020: 18)

LiDAR – Light Detection and Ranging

"LiDAR is a remote sensing technology that measures distance by illuminating a target with a laser and analyzing the reflected light with a detector." (Chai, Nie and Beckder 2020: 24) So it is similar to Radar, but the determining difference is, that the sender doesn't send frequencies, it sends light pulses. (Chai, Nie and Beckder 2020: 24) The major components in this sensor are the laser, the receiver, optics and scanning devices. With this not only the hindrances are recognized, instead the distance is also measured. The measurement principle is direct time off light measurement, which means the measurement time tm is proportional to the distance d: (Chai, Nie and Beckder 2020: 25)

$$d \; = \; \left(\frac{c*t_m}{2} \right)$$

The Doppler-Effect in LiDAR sensors is used for velocity measurement. In this process, requirements and cost are prohibitive to detect the Doppler frequency shift in the light spectrum (Chai, Nie and Beckder 2020: 26). "Differentiation of two or more subsequent distance measurements is used to determine the velocity." (Chai, Nie and Beckder 2020: 26)

$$v_{rel} = \mathrm{d}d \qquad \mathrm{d}t = \lim_{\Delta t \to 0} \Delta d \qquad \Delta t = (d_2 - d_1) / (t_2 - t_1)$$

Ultrasonic Sensor

Another detection of hindrances is the technique of ultrasonic sensors. These use the inverse piezoelectric effect. (Chai, Nie and Beckder 2020: 30) "Through a voltage applied to crystal, the crystal is mechanically deformed, resulting in oscillation." (Chai, Nie and Beckder 2020: 30) And a metal membrane is used to amplify the effect. Sounds in the range of 40 -50 kHz were generated by the oscillation and reflected by objects. The same membrane is used to receive the returned wave. The sensor can only measure very low distances (a range of a few meters), because the distance measurement based on the pulse travel time in low frequencies. So it is used in parking applications. (Chai, Nie and Beckder 2020: 30)

Cameras

A further important qualification of autonomous driving are the cameras in the vehicles. Computer vision utilizes digital cameras in which a digital image sensor converts incoming light into multidimensional signals. (Chai, Nie and Beckder 2020: 23) These special cameras mapped the three-demensional world to a two-demensional image in the imager. Hence the depth dimension is lost during this process, which results in the challenges to be solved by computer vision algorithms. This process is used to extract relevant information for the respective application from the image. (Chai, Nie and Beckder 2020: 24)

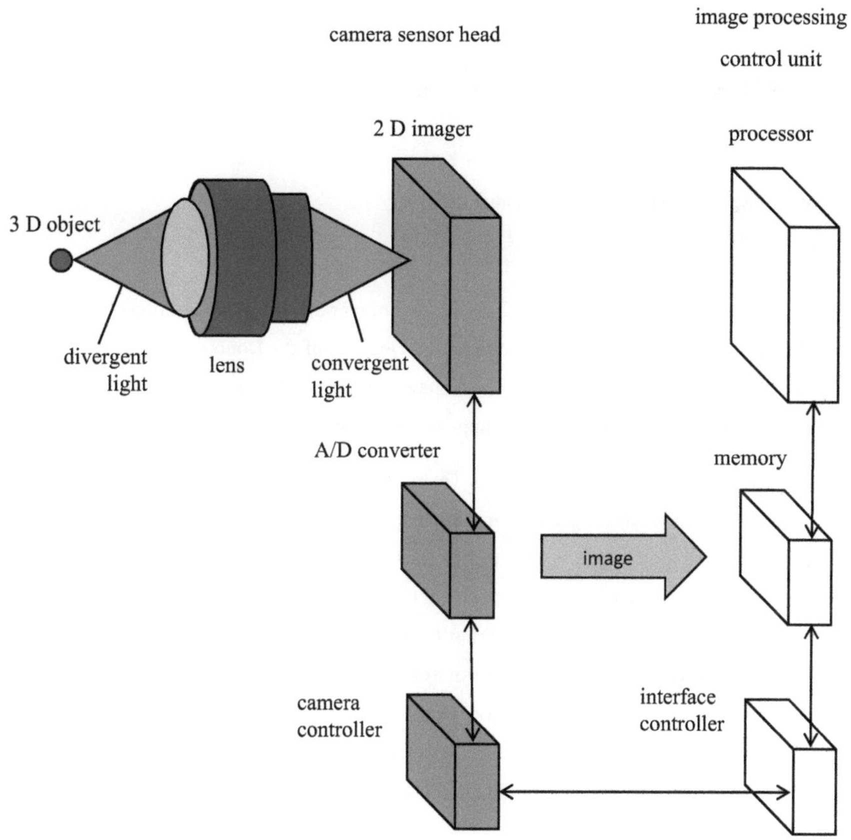

Illustration 9: Principle of an automotive camera (Chai, Nie and Beckder 2020: 23)

4. Handling of Hindrances

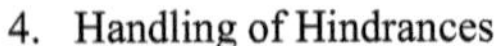

This image has been removed due to copyright issues.

Illustration 10: The Trolley Problem (Feldman 2016)

The Trolley Problem originated in 1967 by the Oxford moral philosopher Phillippa Foot in her paper named "The Problem of Abortion and the Doctrine of the Double Effect" (Panahi 2016). There she describes a scenario in which you have the decision between killing one person or five by switching the railroad track of the approaching tram (trolley) (Beard 2020).

In 1976 the American philosopher Judith J. Thomson wrote a second version of the Trolley Problem to make it more interesting. In this mind game, she tells about the boy George. (Panahi 2016) "George is on a footbridge over the trolley tracks. He knows trolleys, and can see that the one approaching the bridge is out of control. On the track back of the bridge there are five people; the banks are so steep that they will not be able to get off the track in time. George knows that the only way to stop an out-of-control trolley is to drop a very heavy weight into its path. But the only available, sufficiently heavy weight, is a fat man, also watching the trolley from the footbridge. George can shove the fat man onto the track in the path of the trolley, killing the fat man; or he can refrain from doing this, letting the five die." (Panahi 2016)

But what is the connection of the Trolley Problem and Autonomous Driving?

In the Trolley Problem, the human has the decision who the train kills and has to live with the consequences (Panahi 2016). In this situation as well, the human as a driver has the decision in vehicles that we currently know. The decision is if he crashes an animal, a tree or maybe a human, if he has to crash something (external influences are taken into account). But in all cases it is the decision of the human driver and he has to live with the consequences. (Maxmen 2018)

The Trolley Problem as a simplified depiction of an ethnic dilemma leaves room for manoeuvre solutions (Panahi 2016) ,but in combination with artificial intelligence new potentional fatal situations are available (DAVOS 2022). In the new cars, also called autonomous vehicles, the programmed automobile decide (the technical tools taken the external influences into account) in extreme cases who live and who die (Maxmen 2018).

5. The Future of Transport

Everyone is talking about the energy transition (S&P Global 2020) and this phenomenon is the reason why transportation needs to change (Nationalgrid 2021).

The mass of adoption out of combustion engines has led to pollution and city congestion. This is extremely unsustainable for the sake of our planet and human health and happiness. Also, the carbon emissions which arise through cars are now measured high. E. g., in the United Kingdom transporting people and goods, is the largest contributor (responsible for 28 %). (Nationalgrid 2021) "In the US too, transportation is the second-largest source of greenhouse gas emissions (after power generation) and accounts for 70% of all domestic oil consumption." (Nationalgrid 2021)

To reduce these readings and to make lives more comfortable, new solutions have to be created (Reid 2021).

For this, three concepts are signposts for driving the future of transportation: (Galambos 2019)

Smart Technology - "SMART" means self-monitoring, analysis and reporting technology, so this "technology uses artificial intelligence, machine learning and big data analysis to provide cognitive awareness to objects that were in the past considered inanimate" (Bowers 2022) like Internet of Things (IoT) (Bowers 2022).

Electrification - Electricity is indispensable nowadays. This is growing in a continuous process called electrification. (Sympower 2022) Electrification is defined as "the process of replacing technologies that use fossil fuels (coal, oil and natural gas) with technologies that use electricity as a source of energy" (Sympower 2022).

Autonomy - means make your own decision (Collins 2023) So in the automobile industry, it implies the self-determination of vehicles (BMW 2020).

Many predictions of the future of transport can be found in the internet. Here are some examples for traveling in the future: (Galambos 2019)

Hoverbikes

Self-driving taxis

Hyperloop (train tunnel)

… and many more

This image has been removed due to copyright issues.

Illustration 11: The Future Trends In Mobility And Transportation (Marr, 2023)

6. List of Illustrations

Illustration 1: Looser, Devoney. "Horses pull a carriage" 2022. <https://www.wondriumdaily.com/why-the-way-one-traveled-was-important-in-austens-time/>. Web. 09.03.2023.

Illustration 2: Burgess Glasscock, Carl. "The Steam Car" 2011. <https://www.americanautohistory.com/Articles/Article004.htm>. Web. 09.03.2023.

Illustration 3: Woodford, Chris. "The First Car" 2022. <https://www.explainthatstuff.com/historyofcars.html>. Web. 09.03.2023.

Illustration 4: Meriggi, Stefano. "Ford Model T" 2018. <https://www.designindex.org/index/design/ford- model-t.html>. Web. 09.03.2023.

Illustration 5: Greuel, Jorg. "The Trabant" 2018. <https://www.liveabout.com/trabant-built-of-plastic-and-socialism-726030>. Web. 09.03.2023.

Illustration 6: VW Golf AG. "VW Golf" 2022. <https://www.best-selling-cars.com/germany/2022-full-year-germany-best-selling-car-models/#:~:text=The%20VW%20Golf%20has%20been,every%20calendar%20year%20since%201981>. Web. 09.03.2023.

Illustration 7: Team Ackodrive. "Levels of Autonomous Driving" 2022. <https://ackodrive.com/car-guide/autonomous-cars-and-levels-of-autonomous-driving/>. Web. 09.03.2023.

Illustration 8: Schmauch, Julian. "Doppler-Effect" 2022. <https://www.gearnews.de/software-boutique-doppler-effekt-clipping-und-granulareffekt/ >. Web. 02.04.2023.

Illustration 9: Chai, Zhanxiang, Nie, Tianxin and Becker, Jan. *Autonomous Driving Changes the Future*. Singapore: Springer Verlag, 2020, Print.

Illustration 10: Feldman, Brian. "The Trolley Problem Is the Internet's Most Philosophical Meme" 2016. <https://nymag.com/intelligencer/2016/08/trolley-problem-meme-tumblr-philosophy.html>. Web. 10.04.2023.

Illustration 11: Marr, Bernard. " The Future Trends In Mobility And Transportation" 2022. <https://bernardmarr.com/the-future-trends-in-mobility-and-transportation/>. Web. 27.03.2023.

7. Bibliography

American Museum of Natural History. "Trade and Transportation" 2020. <https://www.amnh.org/exhibitions/horse/how-we-shaped-horses-how-horses-shaped-us/trade-and-transportation> . Web. 03.04.2023.

Beard, Matthew. "What's Left To Learn About The Trolley Problem, Philosophy's Most-Meme's Dilemma" 2020. <https://junkee.com/overthinking-it-trolley-problem/280677>. Web. 12.04.2023.

Bekker, Henk. "2022 (Full Year) Germany: Best-Selling Car Models" 2023. <https://www.best-selling-cars.com/germany/2022-full-year-germany-best-selling-car-models/#google_vignette>. Web. 03.04.2023.

Bettex, Morgan. "Explained: the Doppler effect" 2010. <https://news.mit.edu/2010/explained-doppler-0803#:~:text=The%20Doppler%20effect%2C%20or%20Doppler,the%20source%20approaches%20the%20observer>. Web. 08.04.2023.

BMW. "The path to autonomous driving" 2020. <https://www.bmw.com/en/automotive-life/autonomous-driving.html>. Web. 13.04.2023.

Bosh Global. "Artificial Intelligence in autonomous driving" 2018. <https://www.bosch.com/de/stories/kuenstliche-intelligenz-im-auto/>. Web. 26.02.2023.

Bosh Global. "The History of Artificial Intelligence" 2018. <https://www.bosch.com/de/stories/geschichte-der-kuenstlichen-intelligenz/>. Web. 26.02.2023.

Bowers, Kelly. "What Is Smart Technology And What Are Its Benefits?" 2022. <https://rezaid.co.uk/smart-technology-and-its-benefits/>. Web. 13.04.2023.

Burgess Glasscock, Carl. "The Steam Car" 2011. <https://www.americanautohistory.com/Articles/Article004.htm>. Web. 03.04.2023.

Chai, Zhanxiang, Nie, Tianxin and Becker, Jan. *Autonomous Driving Changes the Future.* Singapore: Springer Verlag, 2020, Print.

Collins. "Definition of 'autonomy'" 2023. <https://www.collinsdictionary.com/dictionary/english/autonomy>. Web. 13.04.2023.

DAVOS. "AI's trolley problem can lead us to surprising conclusions" 2022. <https://www.weforum.org/agenda/2022/05/ai-s-trolley-problem-debate-can-lead-us-to-surprising-conclusions/#:~:text=The%20trolley%20problem%20%E2%80%93%20the%20ethical,be%20relegated%20to%20unfeeling%20machines>. Web. 12.04.2023.

Discover. "What are the creative industries?" 2019. <https://discovercreative.careers/students-and-parents/what-are-the-creative-industries/>. Web. 05.04.2023.

Factorial HR. "The Next Generation: Who are They and What do They Need?" 2021. <https://factorialhr.com/blog/next-generation/>. Web. 05.04.2023.

Fait. "What are the level 6 of autonomous vehicles?" 2023. <https://www.faistgroup.com/news/autonomous-vehicles-levels/>. Web. 04.04.2023.

Fandom. "Star Trek: The Next Generation" 2016. <https://memory-alpha.fandom.com/wiki/Star_Trek:_The_Next_Generation>. Web. 05.04.2023.

Galambos, Conrad. "The future of transportation: Where will we go?" 2019. <https://www.geotab.com/blog/future-of-transportation/>. Web. 13.04.2023.

GE Digital. "The Future of the Automotive Industry" 2023. <https://www.ge.com/digital/blog/future-automotive-industry>. Web. 05.04.2023.

Greuel, Jorg. "The Trabant" 2018. <https://www.liveabout.com/trabant-built-of- plastic-and-socialism-726030>. Web. 09.03.2023.

Hammele, Nadine and Kuhnert, Susanne. "The human in the automated vehicle" 2022. <https://www.lpb-bw.de/autonomes-fahren>. Web. 26.02.2023.

Henderson, Tom. "The Doppler Effect" 2015. <https://www.physicsclassroom.com/class/waves/Lesson-3/The-Doppler-Effect>. Web. 08.04.2023.

Hery-Moßmann, Nicole. "How long has there been electricity? The Brief History of Electrification" 2022. <https://praxistipps.chip.de/how-long-has-there-been-electricity-the-brief-history-of-electrification_102258>. Web. 26.02.2023.

History.com. Editors. "Model T" 2019. <https://www.history.com/topics/inventions/model-t>. Web. 03.04.2023.

Infoplease. "Automobile industry" 2022. <https://www.infoplease.com/encyclopedia/social-science/economy/business/automobile-industry>. Web. 05.04.2023.

Lienhard, John H. "The First Automobile?" 2000. <https://www.uh.edu/engines/epi1596.htm#:~:text=The%20earliest%20steam%2Dpowered%20car,by%20springs%20and%20compressed%20air>. Web. 03.04.2023.

Maione, Ines. "All about human and artificial intelligence – differences, strength and weaknesses, human AI and more" 2022. <https://www.clickworker.de/kunden-blog/menschliche-und-kuenstliche-intelligenz-unterschiede-staerken-und-schwaechen-menschliche-ki/#:~:text=Menschliche%20Intelligenz%20beruht%20auf%20der,hat%20diese%20F%C3%A4higkeit%20noch%20nicht>. Web. 26.02.2023.

Maxmen, Amy. "Self-driving car dilemmas reveal that moral choices are not universal" 2018. <https://www.nature.com/articles/d41586-018-07135-0>. Web. 12.04.2023.

Miltimore, John. "The Worst Car Ever: A Brief History of the Trabant" 2019. <https://fee.org/articles/the-worst-car-ever-a-brief-history-of-the-trabant/>. Web. 03.04.2023.

Nakjima, Seio. *The next Generation Automobile Industry as a Creative Industry.* Japan: Eria Discussion Paper Series No. 288, 2019, Print.

Nationalgrid. "The future of transport: driving change in the next 10 years" 2021. <https://www.nationalgrid.com/stories/journey-to-net-zero-stories/future-transport-driving-change-next-10-years>. Web. 13.04.2023.

Noble Wilford, John. "When Humans became Humans" 2022. <https://www.nytimes.com/2002/02/26/science/when-humans-became-human.html>. Web. 03.04.2023.

Panahi, Omid. "Could There Be A Solution To The Trolley Problem" 2016. <https://philosophynow.org/issues/116/Could_There_Be_A_Solution_To_The_Trolle_Problem>. Web. 12.04.2023.

Paoli, Leonardo. "Electric Vehicles" 2022. <https://www.iea.org/reports/electric-vehicles>. Web. 05.04.2023.

Parrish, David. "Creative Industries Definitions" 2015. <https://www.davidparrish.com/creative-industries-definitions/>. Web. 05.04.2023.

Pink, Oliver. "Artificial Intelligence is the key for autonomous driving" Interview by Bosh Global *Artificial Intelligence in autonomous driving* NPR 2018. <https://www.bosch.com/de/stories/kuenstliche-intelligenz-im-auto/>. Web. 26.02.2023.

Poggemann, Malin. "Industrial Revolution: 6 inventions & 6 consequences of industrialization" 2019 <https://www.schreiben.net/article/industrial-revolution-industrialization-7813/>. Web. 26 February 2023.

Reid, Carlton. "Trouble With Predicting Future Of Transportation Is That Today Gets In The Way" 2021 <https://www.forbes.com/sites/carltonreid/2021/09/27/trouble-with-predicting-future-of-transportation-is-that-today-gets-in-the-way/?sh=3e7ebc6d3976>. Web. 13.04.2023.

Shea Choksey, Jessica and Wardlaw, Christian. "Levels of Autonomous Driving, Explained" 2021. <https://www.jdpower.com/cars/shopping-guides/levels-of-autonomous-driving-explained#:~:text=Level%204%20driving%20automation%20technology,geographic%20%20boundaries%20by%20geofencing%20technology>. Web. 04.04.2023.

Snopsys. "The 6 Levels of Vehicle Autonomy Explained" 2023. <https://www.synopsys.com/automotive/autonomous-driving-levels.html> . Web. 04.04.2023.

Stevenson, Jackie. "Why the car industry needs more creativity" 2021. <https://www.shots.net/news/view/why-the-car-industry-needs-more-creativity>. Web. 05.04.2023.

Sympower. "What is Electrification? Definition and Examples" 2022. <https://sympower.net/what-is-electrification-definition-and-examples/>. Web. 13.04.2023.

S&P Global. "What is Energy Transition?" 2021. <https://www.spglobal.com/en/research-insights/articles/what-is-energy-transition>. Web. 13.04.2023.

Team Acko. "Connected Cars: What is it? Features and Benefits" 2023. <https://www.acko.com/car-guide/connected-cars-features-benefits/>. Web. 05.04.2023.

Team Ackodrive. "Levels of Autonomous Driving" 2022. <https://ackodrive.com/car-guide/autonomous-cars-and-levels-of-autonomous- driving/>. Web. 09.03.2023.

The Economic Times. "What is 'Doppler Effect'" 2023 <https://economictimes.indiatimes.com/definition/doppler-effect>. Web. 08.04.2023.

Tremmel, Florian. "Technologies that support and make everyday life easier" 2020. <https://www.silver-tipps.de/technologies-that-support-and-make-everyday-life-easier/>. Web. 26.02.2023.

TWI. "What is an Autonomous Vehicle" 2023. <https://www.twi-global.com/technical-knowledge/faqs/what-is-an-autonomous-vehicle#:~:text=An%20autonomous%20vehicle%2C%20or%20a,ability%20to%20sense%20its%20surroundings >. Web. 05.04.2023.

Tyborski, Roman. "Digitalization and e-mobility: which employees will be in demand automative indusrty in the future" 2020. <https://www.handelsblatt.com/unternehmen/industrie/handelsblatt-auto-gipfel-2020-digitalization-and-e-mobility-which-emplyees-will-be-in-demand-automative-industry-in-the-future/26598330.html>. Web. 26 February 2023.

Verlinden, Neelie. "What comes after Generation Z?" 2021. <https://www.aihr.com/blog/what-comes-after-generation-z/>. Web. 05.04.2023.

Woodford, Chris. "The First Car" 2022. <https://www.explainthatstuff.com/historyofcars.html>. Web. 03.04.2023.